◆

Kecksburg's Sequel?

A Report on the
Mysterious Great Fireball of 1966

Jack Myers

♦

Kecksburg's Sequel?

A Jack O'Lantern Press publication

ISBN: 978-1-0974959-3-1

© **2019 by Jack Myers**
First edition: 2017

All rights reserved. No part of this book may be reproduced in any form or any electronic, mechanical, or other means now known or hereafter invented, including photocopying or recording, or stored in any information storage or retrieval systems without the express written permission of the publisher, except for newspaper, magazine, or other reviewers who wish to quote brief passages in connection with a review.

For inquiries: jackmyers@peoplepc.com

♦♦♦

"I've been convinced for a long time that the flying saucers are real and interplanetary. In other words we are being watched by beings from outer space."

— **Albert M. Chop**, deputy public relations director at NASA

♦♦♦

"Flying saucers are real. Too many good men have seen them that don't have hallucinations."

— **Captain Edward 'Eddie' Rickenbacker**, World War I air ace

♦♦♦

"We all know that UFOs are real. All we need to ask is where do they come from."

— **Astronaut Edgar D. Mitchell**, after his Apollo 14 Moon flight in 1971

♦♦♦

"Maximum security exists concerning the subject of UFOs."

— **CIA Director Allen Dulles**, in 1955.

♦♦♦

"The phenomenon of UFOs is real. I know that there are scientific organizations which study the problem."

— **Mikhail Gorbachev, former Soviet President**, in 1990

♦♦♦

"Of course it is possible that UFO's really do contain aliens as many people believe, and the Government is hushing it up."

— **Professor Stephen Hawking**, University of Cambridge

To
Cecelia Reihl,
for her patience over the years during the researching and writing of these most unusual topics, and

to
MUFON,
for their ceaseless work in getting answers to the world's most perplexing mystery, and

to
Scott Ramsey, Frank Thayer, PhD, and **Stan Gordon**, for encouraging me in publishing this groundbreaking report on a very old yet very new "UFO" case, and

to
Pat and Jack Myers, my parents.

CONTENTS

Preface i

1. April 25th, 1966 1
2. The Fireball Arrives 4
3. Father Knows Best 6
4. Revisiting "The Big Event" 8
5. Meteoric Confusion 10
6. The "Meteor" Impresses 14
7. A Piece of the Rock 18
8. Day and Night 20
9. Pennsylvania's Second Most
   Famous Meteorite 24
10. Official Explanations 27
11. Pennsylvania's #1 Meteorite
    of All Time 30
12. The Kecksburg "Meteorite" Lands 34
13. The Military Covers Up 37
14. Murphy Investigates 41
15. So What Landed in Kecksburg? 43
16. Whatever Happened to the
    Great Fireball of '66? 46
17. Did Something "Big" Land in Quebec? . . . 49
18. Bring in the Helicopter and the
    Search Teams 52
19. The Enigma of Paul Hellyer 56
20. The Governor and the UFOs 62
21. Close Encounters 66
22. UFOs in the Early Days 77
23. Earth vs. the Flying Saucers 82
24. Terror in West Virginia 89

Conclusion 106
About the Author 114

Preface

Something happened on April 25th, 1966 that has never been adequately explained nor fully investigated. This 1966 event occurred little more than four months after the alleged December 9th, 1965 crash landing of an unusual spacecraft or re-entry vehicle just outside the western Pennsylvania town of Kecksburg. Over the intervening years, the situation that unfolded in Kecksburg during the evening hours of 12/9/1965 has come to be known as "Pennsylvania's Roswell."

Readers will note that the Kecksburg incident was precipitated by a fireball that flew over Canada, Michigan, Ohio, and then into western Pennsylvania. Once the fireball "meteor" passed Pittsburgh, it began to engage in a series of inexplicable turns. Decelerating, the object then apparently swooped over Kecksburg and crash-landed in the woods just outside of town where some trees and brush caught fire.

The similarities between the Kecksburg fireball and the subsequent "Great Fireball" of 1966 are inescapable. . . .

Just after 8 p.m. on 4/25/1966 on the East Coast of the United States, a giant fireball appeared suddenly in the clear but darkening skies. Thousands saw the object, and switchboards at newspapers and police stations were jammed with frantic calls about fireballs, meteors, UFOs, rockets, burning aircraft, and yes, flying saucers.

At about the same time the fireball was seen overhead in the Northeast, the Governor of Florida's campaign plane was allegedly being dogged by UFOs in a flight between Orlando and Jacksonville. That UFO sighting, witnessed by virtually everyone aboard the airplane, including the pilots and the governor himself, has never been satisfactorily explained.

Officials in the United States were quick to announce that what witnesses had experienced from the Mid-Atlantic States up to Canada was nothing more than a meteor, a natural phenomenon easily explained. Only problem was, official accounts of the Great Fireball did not sync with a substantial number of the eyewitness reports. Many observers, some with training in such matters, reported an object that moved much lower and much more slowly through the skies than a meteor that had allegedly cruised at an altitude between 25 and 80 miles and at a speed of some 25 miles per second — or faster.

Canadian officials publicly believed that the object blew up and/or crashed in a relatively remote and sparsely populated area of southwestern Quebec. Oddly, even though the best guesstimate of the scientific experts was that the object disintegrated nine miles above the Earth, a helicopter was dispatched to search for the "meteorite" although the chance of recovering a space rock or rocks in the Canadian bush seemed hopelessly futile. Sparing no expense, ground personnel were also sent out on the hunt, including agents to survey residents on what they might have seen.

In the wake of events at Kecksburg, might Canadian officials have been worried some residents of southwestern Quebec had seen things they weren't supposed to see?

Although multiple volunteer first-responders and curious residents of Kecksburg, PA reported seeing a metallic, copper or bronze-colored, Volkswagen-sized craft lying in a gully where the fire in the woods had begun, the official word was that nothing had been recovered in Kecksburg. Nothing. What happened was people had just seen a meteor that authorities had failed to find, if it had ever come down at all.

No one, of course, saw NASA personnel wearing hazmat "moon" suits going into the woods at Kecksburg. No one saw armed military personnel pointing weapons at civilians who refused their orders to leave what was now termed a "quarantined" area. And no one saw a large, tarpaulin-covered object being whisked out of town by the U.S. military on a flatbed truck.

Folks in Kecksburg weren't supposed to see these things.

What things were folks in Quebec not supposed to see?

This report intends to lay out the facts and to, belatedly, open up a new line of inquiry in the UFO community on the "Great Fireball of 1966."

Jack Myers
May, 2017

1 ♦ April 25th, 1966

The modest row house neighborhood from which the author witnessed the Great Fireball.

I was a witness.

The question now, more than a half century later, is what did I actually see?

The event happened in the early evening hours of Monday, April 25th, 1966. I was an 11-year old sixth-grader living in a Southwest Philadelphia row home. After dinner

KECKSBURG'S SEQUEL?

some of my neighborhood buddies and I participated in an informal game of box ball, a close-quarters city game using a tennis ball-sized white rubber "pimple" ball favored in both Boston and Philly. The sun had set, the sky growing a dark bluish purple. Too dark to continue playing. With the enveloping darkness my playmates all returned indoors . . . except for me.

I remained outside on this clear, pleasant evening, lounging on the top step of our front stoop. The Philadelphia Phillies were about to take on their cross-state rival Pirates at Pittsburgh's Forbes Field, the game being televised. Probably leery of being assigned a last-minute chore or, worse, continued homework, I planned on remaining on the stoop until the umpire yelled, "Play ball!"

That's when I saw it.

KECKSBURG'S SEQUEL?

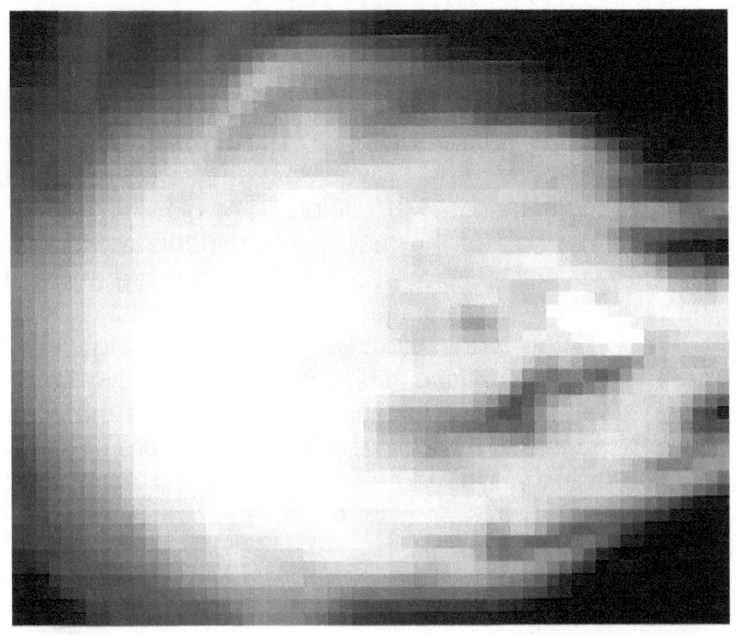

2 ♦ The Fireball Arrives

A large, fiery object appeared suddenly in the sky above the opposing row house rooftops. The fireball, which flew high up in the atmosphere, moved from right to left in front of me as I faced southeast. Multi-colored and glowing, and 50-75% the visual size of the moon, the "meteor" progressed horizontally, taking several seconds to transverse my view of the sky before disappearing behind more rooftops on S. 56th Street. I was mesmerized by the flickering yellow flames that trailed behind this cosmic visitor. No sound accompanied this truly spectacular sight. But as the silent fireball disappeared from view, I was struck by the immediate fear that this large, speeding object might soon come crashing down somewhere in or near Philadelphia — and likely with tragic results.

My approximate view of the fireball's path in 1966 (*Google Maps*).

KECKSBURG'S SEQUEL?

Immediately I darted across our porch and pushed through the front door. My father, seated on the couch, was reading The Philadelphia Evening Bulletin. He was also apparently waiting for the ballgame to begin.

"Dad, Dad . . . a big meteor or fireball of some kind just flew by. I think it's going to crash in Philadelphia!"

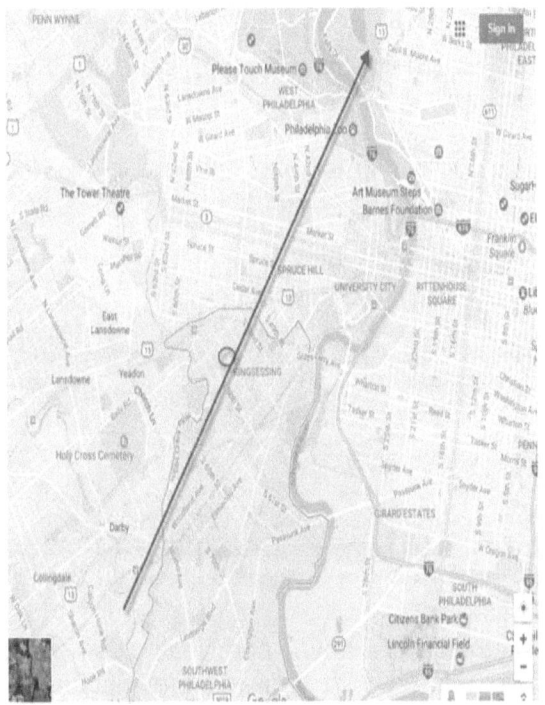

The fiery object's estimated direction based on my Southwest Philadelphia position.

3 ♦ Father Knows Best

My father glanced over the top of The Evening Bulletin's front page. His first years out of Bartram High School had been spent serving in the merchant marines on giant oil tankers that carried Texas crude from the Gulf ports up to the refineries of South Philly. Dad had talked of nights in the company of his fellow bored shipmates, sitting in folding lawn chairs up on deck, cold beverages in hand, swapping stories and watching for shooting stars in the immense ocean darkness.

Merchant marine ship 40 miles southeast of Chesapeake Bay (*National Archives*)

"Those things are never nearly as close as you think," my old man offered with knowing experience. I stood in tense silence for several seconds, listening, waiting for the deafening explosion that never came.

KECKSBURG'S SEQUEL?

Jim Bunning, my father's favorite Phillie, would shut out the Pirates that night. Home run Phenom Dick Allen crushed a fastball from lefty Bob Veal to seal the deal for Philadelphia. Next day the local papers carried the story of the previous night's colossal fireball, vindicating my personal observation of something highly unusual. That the fireball was also spotted by crew members on ships sailing the North Atlantic underscored my father's comment that falling space debris "was never nearly as close as you think."

It was believed that the fireball had flamed out and burnt up somewhere high above the Earth, perhaps even over the ocean. End of story.

And so for the next half century, except on nighttime walks with one eye to the sky, I rarely thought about the great fireball of '66. We moved from Philadelphia in 1970, our beloved neighborhood having crashed and burned by the end of the volatile Sixties.

The author in 1966.

4 ♦ Revisiting "The Big Event"

Recently, after viewing a YouTube video of a California fireball that resembled little of the slow, horizontal flight of the Great Fireball of '66, I went online to see if I could pull up any information on the now half-century old event. To my surprise there were actual photographs, including this fantastic shot by a New York State high school student, Dana DiGeorge, which closely matched my recollection of the slow and steady horizontal "meteor."

Photo taken 4/25/66 by Utica high school freshman Dana DiGeorge (*Life Magazine*).

KECKSBURG'S SEQUEL?

"It looked like a ball that you'd put gasoline on and lit," DiGeorge said, who incredibly was able to snap four photos of the object as it passed.

The only problem was, DiGeorge and a neighborhood buddy noted that the object he photographed was traveling in a westerly direction!

Say what?

5 ♦ Meteoric Confusion

How could a meteor that passed over Philadelphia traveling in a north by north easterly direction wind up in Utica, New York traveling west? Such a flight path made no sense.

DiGeorge was not alone. According to the Utica Daily Press in the article "Bright Object in Sky Startles Area People," Utica resident Rocco DeRocco said he spotted a rotating white light moving west and changing from white to red to green. DeRocco saw the object between 9 and 9:20 p.m., a time confirmed by the fireball photographer Dana DiGeorge.

KECKSBURG'S SEQUEL?

But wait. Many other Utica residents claimed to have seen the mystery object as early as 7 p.m. William C. White of nearby Barneveld and James Brozinski of Utica both said they saw a "blue flash of light" shortly after 7 p.m. Mr. and Mrs. Frank Ervin, driving on Rte. 287 between Barneveld and Prospect, N.Y. described seeing a green object with sparks of fire shooting out the back. Other area residents reported seeing a bright green light, like a skyrocket, shortly after 7 p.m. Many described it as "light blue or green in color, elliptical in shape, and moving rapidly in a northerly direction."

However, April 25th in 1966 was the first day after Daylight Savings Time began. At 7 p.m. it still should have been relatively light outside. Dana DiGeorge's photo of the mystery fireball clearly shows a darkened, nighttime sky.

A spokesperson at Griffiss Air Force Base, after checking with the F.A.A., determined the object to be a "meteorite burning up in the Earth's atmosphere." That was the line that officials repeated continuously on the evening of April 25th, 1966. "Nothing to be worried about, folks, it's just a meteorite."

But Sal Perritano of Utica said the bright light, this meteorite, "revolved" and turn red, white, and green. If this really was a meteorite, it surely was a strange acting one.

KECKSBURG'S SEQUEL?

When officials then received reports of the meteorite being seen in both Albany and Buffalo "simultaneously," they deemed this situation to be "reasonable."

Curtis Hemenway, Director of the Dudley Observatory, must have known something unusual was amiss when he was quoted as saying that the object was probably a meteorite, but "if I were an astronomer on Mars when the Mariner IV went by and someone called to ask me about it, I'd probably have given the same line."

Wink wink and well said, Mr. Hemenway.

Dudley Observatory, Albany (*New York State Archives*)

KECKSBURG'S SEQUEL?

Meanwhile, up in Plattsburgh, N.Y., "to compound the confusion in the sky," persons reported seeing a silvery object giving off blue and red flames as late as 8:45 p.m.

By now the object had flown over much of the northeastern United States, over a reported span of almost 2½ hours, while traveling in several different directions, rotating, and giving off various colors such as white, silver, blue, green, red, yellow, and orange.

To quote from the Air Force file, "there were many conflicting reports on what the object looked like and where it was headed."

6 ♦ The "Meteor" Impresses

Writer Patrick Rowan of Florence, Massachusetts gave what was perhaps my favorite description of the object. "I was 13 years old and playing with a neighbor near my home in Florence when I noticed a star in the southwest — strange since it was still almost daylight. Barely had that registered when a thin white line shot out of the star's left side. The 'star' rapidly brightened and grew, and in an instant it was the 'star' moving, not that white line. Yelling to my friend, we both watched it intensify into a flaming ball the size of a full moon that lumbered sluggishly across the western sky. The head became a hellish churning mass of blazing red eruptions, its teardrop shape leaving a swirling vortex of flame and smoke in its wake. Flaring brightly several times, this qualified as an exploding meteor, or bolide."

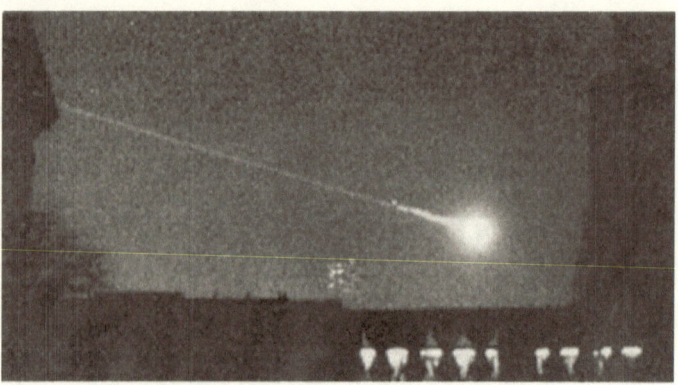

Photo of the 4/25/1966 meteor taken by a Springfield, MA man (*Life* Magazine)

KECKSBURG'S SEQUEL?

Back in Philadelphia, I was not alone in observing the fireball. Decades later, a man named Joe Chmielewski, formerly of Northeast Philadelphia, would post on the Internet about the night he witnessed the "great" fireball. "It had to be huge! From my vantage point in Northeast Philly, as it rolled across the sky, there were only spits of fire on the underside of it. It was only scratching the upper atmosphere. With what looked to be larger than the full moon at a distance of 50 to 75 miles, do the math. An airliner at 60,000 feet is just a small fraction of the apparent size of the moon. This baby was huge! I am not the only one who saw it. We are all not nuts! It was the 60's. It was the Cold War. The powers that were did not want to scare us or maybe didn't want to expose our vulnerabilities. For whatever reason, the reporting of it does not jive with our eyewitness accounts."

What Chmielewski is referring to is that astronomers and other experts such as Dr. Thomas Nicholson from the Hayden Planetarium said that the object was approximately 50-75 miles up when it first touched the upper atmosphere over southeastern, PA . . . was some 25 miles up as it cruised over New York . . . and descended to 9-10 miles in altitude as it crossed the border into Quebec, Canada.

Quebec? Hold on, I thought this thing had flown up to New England and possibly out over the North Atlantic. That appeared to be the direction I saw it speeding above Philadelphia. This sent me searching online for a map of the object's estimated journey through the skies. The following

KECKSBURG'S SEQUEL?

flight path of the 1966 fireball meteor (red line and arrow) is available on the web site **Meteorite Action** at www.fireball.meteorite.free.fr. It shows the fireball beginning somewhere near Chester County, PA and terminating in — yes — southwestern Quebec:

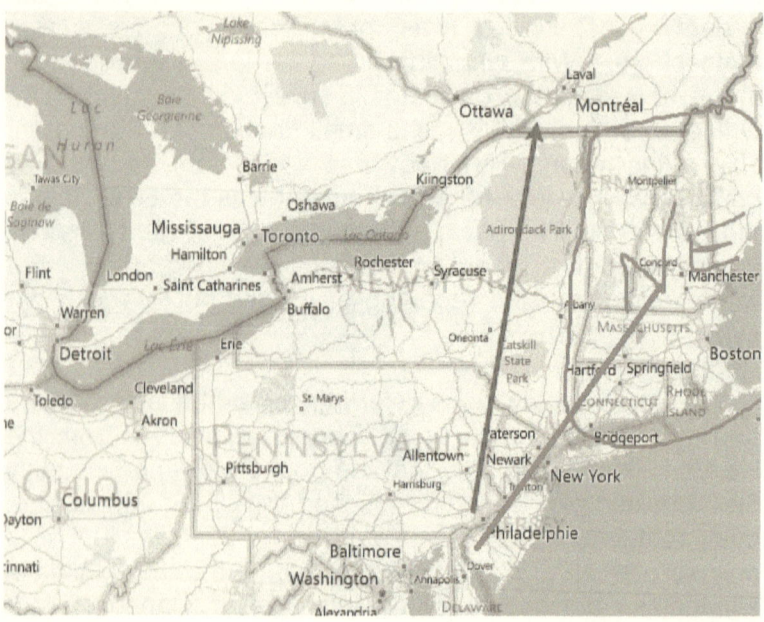

However, the green line shows my perceived flight of the fireball based on my personal observation from my vantage point in Southwest Philadelphia. If the object was indeed at an altitude of 50-75 miles as it entered the atmosphere, and based on the fact the object was high in

KECKSBURG'S SEQUEL?

the sky but definitely in front of me as I looked towards the southeast, then the object may have never flown over Pennsylvania at all, but rather other southern New Jersey.

The web site **Space.com**, in an April 23rd, 2013 article on fireballs, says the 4/25/1966 object in question "passed northward over New Jersey and eastern New York into Canada along a flat trajectory." Canada meaning New Brunswick, perhaps?

And the Federal Aviation Authority announced at 10:23 p.m. on the evening of 4/25/1966 that the meteorite had come down around New Haven, Connecticut.

But most sources, especially later sources, have since claimed that the meteorite "terminated" approximately three miles west of the town of Huntingdon, Quebec . . . and at an estimated altitude of nine miles.

So what gives?

7 ♦ A Piece of the Rock

I decided to check on any other reported sightings in the Philadelphia area, and found an intriguing piece in the Daily Local News of Chester County, PA. This is the approximate spot where the Meteorite Action map shows the meteorite to have first appeared (even though there were sporadic reported sightings farther south in Maryland and North Carolina).

Two suburban youngsters from the Exton area west of Philly (photo from the Daily Local News below) thought they had recovered a piece of the space rock that had fallen onto a hillside. Turns out they didn't after the rock in question was determined by geologists at the local state university to be something other than a meteorite or meteorite fragment.

John Perry and John Green found this small slag-like fragment imbedded in a West Whiteland hillside. They thought it might be from the great fireball meteor of April 25, 1966.

KECKSBURG'S SEQUEL?

Two items in the Daily Local News article immediately bothered me. First, if the meteorite was 50-75 miles up during the early stages of its journey through Earth's orbit, and probably flying over southern New Jersey when it passed Philly, why did two boys from the western Philadelphia exo-burbs believe they had seen a meteorite fall on a nearby hillside?

From my position in Southwest Philadelphia I would have been easily 25 miles closer to the event than these same-aged Exton boys. Several people underneath the meteorite's path thought they saw fragments breaking off and coming down. More than one New York homeowner thought that meteorite debris had fallen on their rooftops! But these youngsters in Exton had not been really all that close.

Second, the Daily Local reported that the "blueish-white meteor which flashed across the skies . . . broke up a baseball game in a West Whiteland cow pasture." This, of course, was the game in which the two 11-year old meteorite hunters had been participating.

8 ♦ Day and Night

Huh? I had been playing box ball on a lit Philadelphia street . . . but my playmates had all quit and gone inside because it had grown too dark to play. But kids in suburban Exton had been playing baseball simultaneously in an unlit cow pasture??

The photos of the meteor from Utica, NY and Springfield, MA bear out my claim that it was far too dark to be playing ball when the meteor passed.

 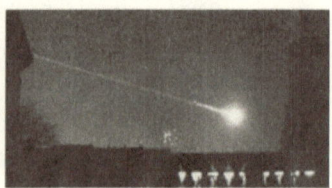

Photos from *Life Magazine*

The Springfield, MA location in daylight
(*Royal Canadian Astronomical Society*)

KECKSBURG'S SEQUEL?

But hold on, you say. Perhaps it was still light in Pennsylvania when the meteorite passed, but dark by the time it flew by upstate New York and Massachusetts?

Nice try . . . but wrong (or at least say the experts).

According to the United States Air Force, astronomers, astrophysicists, and other expects who have investigated the case, the 4/25/1966 meteor's flight lasted approximately 30-31 seconds. That's right, this supposedly "lumbering" fireball covered an estimated several hundred miles in about a half minute. The meteor specialists say the mystery object traveled at 25 to upwards of 30 miles per second. (Nope, not a typo.)

This, of course, would have been quite a surprise to the likes of Dr. Asher Chapman, an examiner for the Federal Aviation Authority no less, who said that he rushed to the window of his Glen Cove, NY home to see the object traveling "about 1,000 feet off the ground, 300 to 400 miles an hour, from south to north."

Some witnesses described an object that flew no more than 100 feet off the ground.

A fighter pilot flying some 50 miles east of Griffiss AFB reported the object had passed "in front of his plane." Not high above it, but in front of it.

KECKSBURG'S SEQUEL?

And one must wonder that if the meteor travelled at upwards of 25 miles per second, how a teenager managed to grab his 1960s style 35mm camera and take four photos. And, how a meteor can travel from South Jersey to Quebec in 31 seconds, yet be sighted by trained observers in Philadelphia at 8:06 and then in New York City fully five minutes later at 8:11. Or how one expert, Dr. Fred Whipple, could estimate the object as football-sized, and another expert, eyewitness Dr. William Bossert, estimate it as nearly football field-sized. Dr. Bossert stated the object was in view over Philly a full 15 seconds.

Oh, and by the way, the Daily Local News of Chester County, PA also wrote that, "Other scientists investigating the recent meteor flight, have indicated that if any part of the object did fall to earth, it would probably be found in Canada or the **New England** states." (emphasis added)

KECKSBURG'S SEQUEL?

I thought about the two boys from Exton, smart and clean-cut Boy Scout types. How could their observation have been so wrong? How could they have been playing baseball in near total darkness and seen a meteor flying over New Jersey that dropped debris on their cow pasture several counties away in PA? Had they just been overexcited? Heck, some people that night claimed to have seen a plane on fire and going down, or even a UFO. Why one caller actually phoned New Jersey's Asbury Park Press to report a passing spaceship, and that the observer could "see a head peering out the porthole." And after all, the Exton/West Whiteland area did border on Downingtown, a southeastern Pennsylvania location that boasts a most extraordinarily famous (or infamous) meteorite . . . one seen by millions.

9 ♦ Pennsylvania's Second Most Famous Meteorite

In 1958, the following meteorite landed in the woods outside of Downingtown, Pennsylvania:

Some kids cruising around town in their hot rods witnessed the fiery object fall just over a wooded hillside, and ran off into the brush to investigate. Soon, an elderly man living in a shack in those same woods found a smoking meteorite in a small crater, and proceeded to poke it with a stick. A slimy, gooey blob came slithering out of the meteorite, jumped onto the man's arm, and eventually killed him.

KECKSBURG'S SEQUEL?

Not finished, the blob continued on devouring the citizens of Downingtown at an alarming rate, while growing ominously in size with each new victim. The teens who had seen the meteorite fall were hip to the danger, but had trouble convincing their parents and the ever-skeptic police that Downingtown was in imminent peril.

I am talking, of course, about the campy 1958 sci-fi/horror movie classic, The Blob, the movie that helped to propel a young actor named Steve McQueen to Hollywood stardom. I have eaten at the Downingtown Diner, "Home of the Blob" which was entirely enveloped by the amorphous space invader at the film's "chilling" climax. I have also watched "The Blob" on the big screen at nearby Phoenixville's Colonial Theater, where 1958 moviegoers were attacked and eaten as the Blob came pouring through the duct work in search of its next meal of human flesh.

KECKSBURG'S SEQUEL?

So young John Green and John Perry of West Whiteland can logically be excused for getting overexcited, having almost undoubtedly seen The Blob, Chester County's legendary teen horror flick. But that still doesn't fully explain why the boys thought a faraway meteor had flown directly overhead — or how they came to be playing baseball in a cow pasture after dark. . . .

Unless . . . unless the meteor had first flown over Chester County earlier in the evening, then perhaps went out to sea, and later turned back high over South Jersey when I came to see it shortly after dark?

But wait. Meteors don't make turns, do they? And if they do make turns, then they really aren't meteors, correct? And when is a meteor, well, not a meteor?

The Blob, 1958 (*Paramount Pictures* and *Jack H. Harris*)

10 ♦ Official Explanations

According to Project Blue Book files from 1966, General Electric electrical engineer, J. Russell Clark of Whitesboro, NY said he and his son, Don, 15, saw the object break up "within 50 miles." The father said, "I'm sure it wasn't a meteorite. It was too low and too flat a trajectory." He estimated its altitude at 1,000 to 5,000 feet. Clark said it might have been some "space hardware" re-entering the atmosphere.

Clark, of course, did not mention whether he suspected this flaming "space hardware" to be of terrestrial or extraterrestrial origin.

A Brown University professor, who had not witnessed the event but who had been undoubtedly called on by authorities to calm the public, issued the following straight-faced statement. Project Blue Book noted that Dr. Charles H. Smiley, a Brown astronomer, "smiled the smile which only men who have demolished a flying saucer smile" and pronounced the object in question to be a meteor. With that happy face official introduction, the esteemed academic offered this leaned opinion:

"As a scientist I'm inclined to say there has been no evidence they (the meteors?) come from anywhere outside the Earth. All I've seen and heard about are natural phenomena not understood by the public."

KECKSBURG'S SEQUEL?

Really? Meteors do not come from anywhere outside the Earth? And to think I passed on applying to the Ivy League just because my grades weren't up to snuff.

Dr. Smiley then went on to blame flying saucer sightings on the press, TV publicity, and the human desire for recognition.

Any thoughtful observer here would question the need for authorities to trot out a Dr. Smiley and have him "calm the public" after such an event. What might have had government officials so nervous here? And why were they

KECKSBURG'S SEQUEL?

so frightened on the subject of flying saucers and extraterrestrial visitors . . . something they had repeatedly denied even existed?

A look at the case involving Pennsylvania's most famous meteorite of all can perhaps provide some much needed clues.

11 ♦ Pennsylvania's #1 Meteorite of All Time

KECKSBURG'S SEQUEL?

In the early evening hours of December 9, 1965, another "fireball" entered the Earth's atmosphere, this time somewhere over Canada. This object, appearing barely four months before the Great Fireball of '66, was tracked by military radar as it cruised into U.S. air space over Michigan, Indiana, Ohio, and then on into western Pennsylvania.

It was when the "fireball" passed near Pittsburgh that things started to get weird.

Really weird.

The so-called meteor began to slow down, decelerating to 300 mph or less.

And it also began to make a series of turns.

Witness Bill Bulebush, who was working on a CB radio in his Corvair, happened to catch sight of a blueish-white light passing overhead. Bulebush said it appeared the object next tried to fly over a mountain, couldn't reach the necessary altitude, and then reversed course.

Randy Overly, a local youngster playing in a field with friends, watched the mystery object approach from a distance, and then observed it cruise almost directly overhead at a height of about 200 feet. Overly described the "meteor" as being acorn shaped, orange and brownish in color. It made a distinct hissing noise as it passed, and was surrounded by some sort of "vapor." There was fire coming

KECKSBURG'S SEQUEL?

out the back of the object, and Overly could distinctly see greenish, reddish yellow flames.

The witness would years later state, as an adult, that the "meteor" wasn't really a meteor at all. "It was a constructed thing," Overly insisted. "It had smooth edges and lines."

Pennsylvania researcher Stan Gordon, who has been investigating "Pennsylvania's Roswell" for decades, is shown on the previous page standing in front of a mockup of the fireball "meteor" that wasn't a meteor. Gordon has painstakingly documented how the object next crashed (or landed, depending on the point of view) in the woods just outside of the sleepy western Pennsylvania hamlet of Kecksburg.

The details about "The Other Roswell" can be found in Gordon's first-rate documentary **Kecksburg: The Untold Story**.

KECKSBURG'S SEQUEL?

www.stangordon.info

12 ♦ The Kecksburg "Meteorite" Lands

While no one apparently saw the crash (or crash landing), smoke was soon spotted coming from a wooded area on the outskirts of town. With phoned in reports of some kind of a flying craft apparently in distress, volunteer firefighters from multiple municipalities were dispatched to the scene. What they expected to find was a crashed and damaged airplane. What they stumbled upon partially burrowed into a small wash or gully in a local hollow was entirely something unexpected.

Volunteer firefighter Jim Romansky was one of the first to arrive at the "impact site." Romansky describes a large, metallic, acorn-shaped object partially submerged in a gully. The craft was large enough, according to Romansky, for a grown man to stand up and move around in. There was strange writing like Egyptian hieroglyphics that encircled the object's "bumper" near its base. Copper or bronze in color, the craft had no portholes, windows, doors, wings, tail assembly, seams, or rivets. It appeared to Romansky, and other witnesses, to have been fashioned out of a single molded piece of metal.

ARMY ROPES OFF AREA—
'Unidentified Flying Object' Falls Near Kecksburg

KECKSBURG'S SEQUEL?

Several other locals made their way down to the impact site, including other firefighters, local officials, curious neighbors, and a smattering of pesky teenagers. One person early on the scene was veteran Greensburg WHJB news director John Murphy. Murphy, who came equipped to do a story, immediately notified state police of the incident, and then began taking copious notes, photographs, and also taping the first-hand statements of the various witnesses.

Soon, however, convoys of military personnel began descending on the relatively obscure and out-of-the-way village in Westmoreland County. Some units came in uniform, especially those arriving in personnel carriers, while others were plainclothes men dressed in overcoats against the December cold. Some were reportedly Army, while others were Air Force. The military people immediately cordoned off the impact site and ordered everyone, including Murphy and the volunteer responders, out of the woods. They could be seen searching the woods with flashlights. Many of the men stationed on the road or around the woods wore helmets and carried rifles or side arms.

Folks naturally asked if they could go into the woods, and were told that the area was now restricted and off limits. They were informed the area had possibly been contaminated by radiation and was being quarantined. Those who persisted had weapons pointed at them and were threatened with arrest. One pushy teen was warned he was about to have his prized automobile confiscated.

KECKSBURG'S SEQUEL?

Kecksburg, it appeared, had been placed under a temporary state of martial law.

Pennsylvania State fire marshal Carl Metz, one of the officials summoned to Kecksburg partially based on Murphy's notification of the authorities, went down to the gully to assess the situation and to speak with the ranking military officers in charge. After a while, Metz returned from the woods.

Murphy, who had been waiting impatiently nearby the cordoned off "impact site," approached Metz when he returned.

"So Carl, what do we have down there?"

"I'm not sure," replied the puzzled and apparently distracted state fire marshal.

When pressed by Murphy for more information, Metz answered tersely, "You'll have to ask the military."

13 ♦ The Military Covers Up

Murphy tried doing just that — he asked the military — and for his trouble was relieved of his notes, camera, and the audio tape he had made of witnesses interviews. Reportedly Murphy had a roll of film in his pocket that was not confiscated, but this has never been substantiated.

Officially, there were only three members of the military sent to Kecksburg. But according to townspeople and other witnesses, scores arrived in separate detachments. They took over the local station house and fire department building, and also commandeered a farmhouse that stood close by the impact site. Members of the family that lived in the farmhouse recalled telephone calls being made to NASA. The calls did not appear, however, on any of the family's subsequent phone bills.

Murphy's radio station office manager, Mabel Mazza, remembered taking calls from several governmental agencies, including the Air Force and the Pentagon. Most were seeking directions to the Kecksburg impact site.

Later that same night, NASA personnel dressed in white hazmat coveralls that some locals referred to as "moon suits" were seen descending down into the wooded hollow. They were soon seen carrying out smallish boxes, no more than four to five feet long, contents unknown.

KECKSBURG'S SEQUEL?

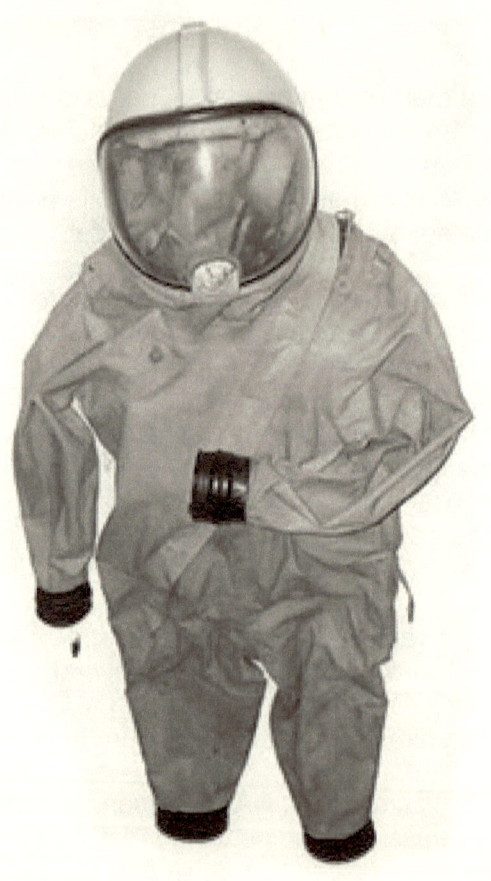

Eventually, very late that night, the military was able to hoist the downed object onto a large, flatbed truck and remove the object from the impact area. Kecksburg

KECKSBURG'S SEQUEL?

residents reportedly witnessed military jeeps front and back, red lights flashing, escorting the flatbed truck down "Snake Hill" and out of town. The military escort roared out of Westmoreland County at a high rate of speed.

Although the recovered object was completely covered with a tarpaulin, the protective material had been drawn taut and the "acorn" or "bell" shape of the device could plainly be seen by all.

"And they weren't stopping for nothing," noted one alleged onlooker. "I've you would've stepped in the way they would've been scraping you off the road with a putty knife."

Anecdotal evidence suggests the flatbed truck headed west out of Pennsylvania and into Ohio. The flying "acorn" was temporarily taken to an Air Force base near Columbus, and then later transported to its ultimate destination — Wright Patterson Air Force Base in Dayton.

Officials soon announced that what had happened over the skies of southwestern Pennsylvania had been little more than a fireball type meteor. And a search of the Kecksburg woods had found . . . nothing.

That's correct. Zip, zilch, nada, not a thing.

KECKSBURG'S SEQUEL?

The Greensburg Tribune Review reported that U.S. Army engineers and civilian scientists gave statements suggesting that absolutely nothing had been found.

14 ♦ Murphy Investigates

Undaunted, WHJB news director John Murphy starting working on what he felt would be, by far, the biggest news story of his career. He re-interviewed witnesses and dug deeply for more eye-opening material. Murphy was going to do a radio show on the "crash" and break the case wide open. Something highly unusual had landed in the woods . . . and people needed to know about it. The truth needed to come out.

But then, a funny thing happened. Witnesses began calling Murphy to recant. They had thought better of it, they claimed, and didn't want what they had said to be on the air. They were afraid the military and the state police might "be mad at them."

KECKSBURG'S SEQUEL?

Next, radio station employees witnessed some official looking men in suits visit Murphy in his office. Afterwards, Murphy seemed agitated. He stopped working on the crash story, and he also stopped talking about it. The planned radio show aired, but it was a watered down, toothless version of what John Murphy had originally envisioned. This story had been Murphy's baby . . . and now suddenly it was an orphan.

"That wasn't the John I knew," offered Murphy's wife while on camera in Stan Gordon's Kecksburg documentary. "They got to him."

Years later, while on vacation in California, Murphy was inexplicably killed by a hit-and-run driver. The case was never solved. Many, including his wife, believed Murphy's death to be suspicious, and somehow related to the Kecksburg UFO incident.

Ever since, it has been Stan Gordon's job to run with the story. For details visit www.stangordon.info.

15 ♦ So What Landed at Kecksburg?

When volunteer firefighter Jim Romansky and a fellow first responder were "relieved" of their duty at the Kecksburg impact site, one of the men naturally asked the arriving military personnel, "What is it?"

"It's a meteorite," came the reply.

The incredulous volunteers waited to get out of earshot before letting out a round of nervous laughter.

"That's no meteorite!" one man exclaimed to the other.

"Nope, you're right. That was no meteorite."

So, if not a meteorite, then what was it?

A foreign satellite of some kind? Russian or Chinese perhaps? Or even an exotic new American prototype?

An intriguing possibility recently bandied about was that the "acorn" was a later version of Nazi Germany's storied Wunderwaffe or "wonder weapon." As the legend goes, by 1944-45, the Nazis were feverishly working on an amazing new weapon that would turn the tide of the war for Hitler and his Third Reich. Somewhere at a secure base just east of Germany, top Nazi scientists were developing an anti-gravity flying machine known as Die Glocke or "the Bell."

KECKSBURG'S SEQUEL?

The strange device was shaped like a bell, and disappeared in 1945 as Berlin fell.

After World War II, many Nazi scientists were brought to the U.S. to work on the American rocketry and, eventually, space programs. This effort was known, innocuously, as Operation Paperclip. Some have speculated that the plans for the Nazi bell may have survived with German scientists, resulting in a secret U.S. prototype of Die Glocke having to

KECKSBURG'S SEQUEL?

do an emergency landing in Kecksburg. Hence, the scores of military that descended upon the impact site in an unbelievably rapid fashion.

And if not a satellite or some secret, experimental new terrestrial craft? Then, we are talking UFO, flying saucer, and extraterrestrial flying machine, correct?

Oh, but right, they don't officially exist. Except that Roswell insiders such as Major Jesse Marcel and Lt. Col. Philip J. Corso already spilled the beans before their deaths.

Okay, so back to the Great Fireball of April 25th, 1966. What took place on that date?

Fireball meteor . . . or something else??

16 ♦ Whatever Happened to the Great Fireball of '66?

Well, according to the eyewitnesses, the Great Fireball of 1966 started out much the same as the Kecksburg fireball that preceded it by less than five months. Bright fireball streaking through the early evening sky, seen by people in multiple northern states.

But, of course, most fireballs are just that . . . fireballs.

Was there anything unusual about the 1966 fireball meteor, besides its exceptional brightness and flat trajectory? Or was it just a big, burning hunk of space rock?

Yes, what happened to the Fireball of '66?

According to the August, 1967 Journal of the Royal Astronomical Society of Canada, the "meteor" terminated flight some nine miles above the Earth at a position three miles west of the town of Huntingdon in southwestern Quebec. By "terminated" it is meant that the object exploded during flight into a shower of pebbles and dust.

KECKSBURG'S SEQUEL?

The Canadian astronomical journal relates how a farmer, standing on a high point of land about one mile north of the Canadian-U.S. border, was facing south with an unobstructed view and low horizon. The farmer saw the fireball appear above some trees on the distant horizon. It then appeared to rise vertically and pass directly overhead. The farmer turned around to see the fireball disappear "a little to the north."

"It was now apparent that the fireball had started far to the south but definitely ended in Canada where the major meteorite fragments should be found," explained the journal.

"Initial data placed it over the region south of Ottawa-Montreal and heading northeast towards St. Jean, Quebec. . . . It was apparent that the fireball had travelled from south to north and passed near but to the west of Huntingdon."

It is important to note that none of the meteorite's fragments have ever been found, even though the Royal Astronomical Society maintains that it was "possible to determine accurately the end-point" based on the numerous credible eyewitness accounts. Near the end-point observers related hearing "hissing and crackling noises during passage of the fireball."

Many thought it was . . . an approaching aircraft, same as in Kecksburg.

KECKSBURG'S SEQUEL?

Reportedly, the fireball "burst and disintegrated into an elongated cluster of fragments of varying sizes, colors, intensities, and duration." Also, observers relatively close to the end-point were consistent in reporting two luminous fragments. "Many observers from the region to the northwest were persistent in reporting either a black ball or dark objects continuing beyond the point where the luminosity ended."

A. A. Griffin, with the Dominion Observatory in Ottawa, wrote that he believed the meteor travelled northward to the St. Lawrence River, and that its "luminous trail ended over the area a few miles northwest of Huntingdon, Quebec."

17 ♦ Did Something "Big" Land in Quebec?

Meteor Crater near Winslow, Arizona (NASA)

B.A. McIntosh and J.A.V. Douglas wrote in their August, 1967 journal that "observers near the end position described the fireball as a round ball which grew in dimensions as it approached and shot off sparks. These observers were not able to see the disintegration and trail of fragments behind the head of the fireball."

KECKSBURG'S SEQUEL?

So, the object appeared to be big and round before the explosion. And after?

McIntosh and Douglas reported that, toward the end of the illuminated path, "Observers relatively close to the endpoint were consistent in reporting two luminous fragments." **Also, that two independent observers described the objects seen as "being as big as an automobile."**

It should be noted here that some of the witnesses in Kecksburg described the metallic "acorn" as being as big — or bigger — than a Volkswagen automobile.

McIntosh and Douglas pooh-poohed the idea of an object or objects the size of automobiles surviving the meteor's flight. They added that these independent observers "must have unconsciously corrected for distance, since most people indicated size in relation to small objects such as a golf ball."

So, no bother looking for golf ball-sized and smaller fragments in the rugged moose pasture of Quebec, correct? The witnesses hadn't seen what they thought they saw.

And especially not if, as McIntosh and Douglas wrote, that fragments may have been scattered for miles and miles beneath the trajectory of the meteorite . . . even before it crossed into Canadian airspace. ". . . it appears likely that part of the April 25 mass came down in the rugged terrain of the Adirondack Mountains."

KECKSBURG'S SEQUEL?

So, end of story, case closed? Well, not so fast. There is more. . . .

18 ♦ Bring in the Helicopter and the Search Teams

As per McIntosh and Douglas, "A search by helicopter was carried out in the area under the trajectory from the Canada-U.S. border to the south shore of the St. Lawrence River. Interview teams on the ground covered this area and also the region under the extended trajectory between the St. Lawrence and Ottawa Rivers."

Say what? A helicopter and search teams to look over an extended swath of rugged terrain in late April for a space rock that likely disintegrated into a cloud of dust, pebbles, and fragments no larger than the size of golf balls? Who was paying for all that? And why?

Makes no sense . . . unless they were looking for an object or objects much larger, and of much higher value and importance than a disintegrated space rock. Was the almost instantaneous official media blitz to have this object labeled as a "fireball" a cover for a much bigger story?

A. A. Griffin reported that ". . . staff members of the Dominion Observatory went to this area to get information which might lead to the recovery of a meteorite. In cooperation with representatives of the Geological Survey of Canada and the National Research Council (McIntosh and Douglas 1967) local inhabitants were interviewed. . . . A few days later this terrain, of freshly ploughed fields, pasture,

KECKSBURG'S SEQUEL?

and swamp was thoroughly searched by helicopter. From our cruising height and against this background we stood a good chance of spotting a meteorite a foot or more in diameter. The search was not successful but some farmer, alerted by interviews, may yet find this meteorite."

Really? Most space rocks, especially from stony meteorites, look just like Earth rocks, especially from a passing helicopter. And northern latitudes are littered by debris from boulders pulverized by advancing and retreating mile-high glaciers during the last Ice Age.

Some 50+ years later, the remnants of the Great Fireball of '66 have still not been officially found. Not a speck.

Now, if you were looking for an object as big, or bigger, than a Volkswagen, a copper or bronze colored one-piece

KECKSBURG'S SEQUEL?

metallic acorn-shaped re-entry vehicle, then I would say the helicopter fly-over might possibly have yielded some results.

And perhaps it did. . . .

Which begs the question, were the local inhabitants being interviewed about what they saw — or were they being debriefed about things they maybe shouldn't have seen?

The report by McIntosh and Douglas tends to contradict A.A. Griffin's optimism about the chances of finding a meteorite in southwestern Quebec:

"The anticlimax of so spectacular an event is surely the failure to recover meteorites. Although there is no direct evidence to show that meteorites did fall, it is more probable that none have been found due to one or both of the following:

1. First, much of the possible fall area is either sparsely populated or unfavorable terrain for recovery. This includes the 5-mile stretch of open water of Lake St. Francis which begins only 7 miles beyond the end-point.

2. Secondly, although most large meteorite falls are accompanied by dust clouds, the combination of evidence of the early explosions and of the cinder cloud which continued beyond the end-point may

KECKSBURG'S SEQUEL?

indicate that the meteorite disintegrated into relatively small fragments.

So, the chance of recovering tiny, inconspicuous-looking fragments over a vast terrain by means of using a search helicopter was astronomically small. Unless . . .

. . . unless they weren't looking for small fragments — or even meteorite(s).

But is there any evidence that the Canadian authorities and/or military did recover something of an unusual nature from the Great Fireball of 1966?

The secret may yet lie with a controversial Canadian public figure, one Mr. Paul Hellyer.

19 ♦ The Enigma of Paul Hellyer

Paul Theodore Hellyer is a Canadian politician, engineer, writer, and commentator who has had a long and varied career. He is the longest serving member of the Privy Council of Canada just ahead of Prince Philip. For those of you who aren't "royal watchers," Prince Philip, the Duke of Edinburgh, happens to be the husband of Queen Elizabeth II of Great Britain. Suffice it to say that Mr. Hellyer moves in some extremely rarified circles.

Just over a decade ago, in September 2005, Hellyer created something of an uproar when he publicly announced his belief in the existence of UFOs. According to CNN,

KECKSBURG'S SEQUEL?

"Former Canadian Minister of Defense & Deputy Prime Minister, Hon. Paul Hellyer says, 'UFOs are as real as the airplanes flying over your head.'"

Then, in 2007, the Ottawa Citizen reported Hellyer had demanded that world governments disclose alien technology so as to help solve the global problem of climate change.

In perhaps his most eye-popping announcement to date, Hellyer, in 2014, during an interview with the RT Network (formerly Russia Today), was quoted as saying that "at least four species of aliens have been visiting Earth for thousands of years . . . and they don't believe we are good stewards of our planet."

Now, perhaps Hellyer is just some poor, old duffer who has finally snapped his cap. In fact, Yahoo News reacted to Hellyer's "alien visitation" statement by commenting this was "sadly hard to take seriously."

But nevertheless, Hellyer apparently remains, at the time of this writing, a member in good standing of the Canadian establishment. According to a 2011 article in U.K.'s Daily Mail, Hellyer is on an advisory body to the Queen, works as an environmental campaigner, and is credited with integrating Canada's armed forces.

Other than his views on UFOs and alien visitation, Mr. Hellyer is quite obviously taken very seriously. Meanwhile, the Canadian's web site boasts that he is the first person of

KECKSBURG'S SEQUEL?

cabinet rank in the G8 group of countries to state unequivocally "UFO's are as real as the airplanes flying overhead."

The former Canadian Defense Minister readily agrees, according to the Daily Mail, that he would in all likelihood be fired for his unorthodox beliefs if he was still Canada's Minister of National Defense today.

So how did the Hon. Paul Hellyer acquire such a burning passion regarding the extremely touchy, normally off-limits subjects of UFOs in our skies and the alien visitation of Earth? Why does he steadfastly claim that evidence concerning UFOs is the "greatest and most successful cover up in the history of the world," according to Exopolitics.org?

Hellyer is adamant that he, his wife, and some friends all witnessed a UFO event years before, but that he was quick to discount the incident at the time. Rather, maintains Exopolitics.org, Hellyer's position on UFOs dramatically changed after watching the late Peter Jennings' documentary special, "Seeing is Believing" in February 2005. Afterwards, Hellyer decided to read a book that had been idly sitting on his book shelf for two years, Philip Corso's <u>The Day After Roswell</u>. The book sparked intense interest for Hellyer as Col. Corso named real people, institutions and events that could be checked. Hellyer decided to confirm whether Corso's book was real or a "work of fiction." He contacted a retired United States Air Force

KECKSBURG'S SEQUEL?

General and spoke to him directly to verify Corso's claims. The unnamed General simply said: "every word is true and more."

Mr. Hellyer, of course, denies he gained any secret knowledge of UFOs in his position as Minister of Defense in Canada. But Hellyer may not be at liberty to reveal specifically what he learned as Defense Minister because that would be in direct violation of laws preventing the dissemination of classified, top secret information. Hellyer would have taken an oath to uphold those very laws. So, ironically, Mr. Hellyer himself may be an unwitting, duly

KECKSBURG'S SEQUEL?

sworn party to what he calls the "greatest and most successful cover up in the history of the world." In order to continue earning the title of being "the Honorable Paul Hellyer" the former minister is duty bound to keep certain state secrets.

What we do know is that the Hon. Paul Hellyer served as Canada's Minister of National Defense from April 22, 1963 until September 18, 1967. Yes, the Great Fireball of 1966, which allegedly came down in Quebec, would have landed during Mr. Hellyer's watch. If anyone was privy to classified information concerning the "object" or "objects" that fell on Canada during the evening of April 25th, 1966, it would be Paul Hellyer.

Does Hellyer's fascination with UFOs and aliens stem from something the Canadian military recovered, out of civilian sight, in the "sparsely populated, unfavorable terrain" of southwestern Quebec?

Hellyer's interest in UFOs and aliens would appear to extend much further back than he would care to admit. On June 3, 1967, as Canadian Minister of National Defense, Hellyer flew in by helicopter to officially inaugurate a UFO landing pad in St. Paul, Alberta. The town had built it as its Canadian Centennial celebration project, and as a symbol of keeping space free from human warfare.

KECKSBURG'S SEQUEL?

The UFO landing pad in St. Paul, Alberta (*Town of St. Paul web site*).

The Paul Hellyer story, circumstantial by nature, provides little in the way of solid answers to the mystery of the Great Fireball of '66. If anything it only poses more questions. So, I decided to check back on the date of April 25th, 1966. Did anything else unusual or peculiar happen on or around that date?

20 ♦ The Governor and the UFOs

> **'Mysterious UFO' Escorts Burns Plane**
>
> ★ ★ ★
>
> **Meteorite Flares Over Northeast**
>
> Times Staff Writer
>
> ABOARD THE BURNS CONVAIR — It's been a long political campaign but last night Gov. Haydon Burns, four cynical newspapermen, Mrs. Burns and six Burns staff members saw two spots of light that looked suspiciously like an un-

On the very same evening a giant fireball flew over the northeastern portion of the United States, and then allegedly disappeared over Canada, another strange occurrence was taking place in the skies of Florida at approximately the same hour.

The plane carrying Florida Governor Haydon Burns, who was returning from a campaign trip to Orlando, was reportedly dogged by two UFOs.

KECKSBURG'S SEQUEL?

That's correct . . . both a giant fireball and two UFOs were seen along the United States' eastern coastline on the same evening at approximately the same time.

A pair of mysterious lights appeared shortly after the Governor's aircraft took off, and supposedly trailed the plane for a distance of at least 40 miles. Witnesses to the incident included Governor Burns' pilots, a Florida Highway Patrol captain, six of Burns' staff members, the Governor's wife, and four newspaper reporters along for the eventful ride.

Reporter John Meiklejohn of the St. Petersburg Times, himself a witness, wrote that while looking out one of the airplane's windows he observed two round lights that appeared to be trailing the plane and maintaining speed.

Burns' campaign plane was flying at an estimated altitude of 6,000 feet at 230 mph.

Then, at approximately 8:52 p.m., Burns reportedly blurted out, "It's a UFO!"

The startled Governor next ordered his pilots to turn into the objects. However, at almost the same instant Burns had issued his order, the St. Petersburg Times wrote that "the lights appeared to rise and then disappeared as though the electricity had been switched off."

KECKSBURG'S SEQUEL?

Burns is said to have later made jokes about the sighting, boasting that "I told you my campaign would be out of this world!"

The UFO incident, however, did not help to prevent Burns from losing in a runoff election to his Democratic rival.

Some observers would later attempt to connect the 4/25/1966 Florida UFO incident to the almost simultaneous

KECKSBURG'S SEQUEL?

appearance of the Great Fireball of 1966, but no direct link was ever found.

It should be noted that no astronomers, astrophysicists, or other experts were apparently called in to explain to Governor Burns, his wife, the Florida state police, the Governor's staff members, the news reporters, or the Convair's pilots that what they had actually witnessed was a meteor, the planet Venus, or perhaps the Moon over Miami.

The Florida UFO incident of April 25th, 1966 remains officially unexplained to this day.

Incidentally, the St. Petersburg Times, Ocala Star-Banner, and other Florida newspapers were citing reports at the time that the meteorite up north had landed in Long Island Sound. Not in New Haven, New York's Adirondacks, or in Quebec, but off Long Island, NY.

21 ♦ Close Encounters

Intrigued by the Florida governor's plane being involved in a multiple-witness UFO incident at the very same time as the Great Fireball of '66 was streaking across U.S. skies, I searched for other unusual incidents that may have occurred within in that same time frame. My search did not take long.

Eight days before the Great Fireball appeared over eastern Pennsylvania, police in Ohio and western Pennsylvania experienced an encounter that has become the stuff of legend.

KECKSBURG'S SEQUEL?

In the early morning hours of Sunday, April 17th, officers of the Portage County, Ohio Sheriff's Department would chase a structured, low-flying UFO for some 86 miles across state lines into Pennsylvania.

Along the extended chase route, the UFO played a cat-and-mouse game with pursuing law officers. Also on that mainly east-bound route, cops from other jurisdictions would see the brilliant object and join in on the wild and woolly chase, which ultimately proved fruitless.

Things began at 5 a.m. with Deputy Sheriff Dale Spaur and Mounted Deputy Wilbur "Barney" Neff out investigating an apparent abandoned vehicle on the highway. A call had come in minutes before about a woman who had phoned to report a bright light crossing the sky above her house. Spaur and Neff were more immediately concerned with the abandoned vehicle. That quickly changed.

They Tailed a Saucer 86 Miles!

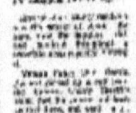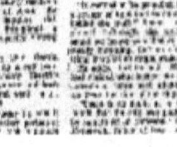

In Deputy Sheriff Spaur's words, "I always look behind me so no one can come up behind me. And when I look in this wooded area behind us, I saw this thing. At the time it

KECKSBURG'S SEQUEL?

was coming up . . . to about tree-top level. I'd say about one hundred feet. It started moving towards us. . . . As it came over the trees, I looked at Barney (Deputy Neff) and he was still watching the car . . . and he didn't say nothing and the thing kept getting brighter and the area started to get light . . . I told him to look over his shoulder and he did.

"He just stood there with his mouth open for a minute, as bright as it was, and he looked down. And I started looking down, and I looked at my hands and my clothes weren't burning or anything, when it stopped right over top of us. The only thing, the only sound in the whole area was a hum, like a transformer being loaded or an overloaded transformer when it changes.

"I was petrified, and so I moved my right foot, and everything seemed to work alright. And evidently he (Deputy Neff) made the same decision I did, to get something between me and it, or us and it. . . . So we both went for the car, we got in the car. . . ."

Spaur also described the object as being "very bright . . . it'd make your eyes water."

As the officers watched, the object drifted east, then stopped. Spaur used the opportunity to radio their dispatcher. Their sergeant got on the radio and told the officers to follow the object, and meanwhile they'd try to get a photo unit on scene.

KECKSBURG'S SEQUEL?

With the object on the move, Spaur and Neff turned south on Rte. 183, then east on Rte. 224.

"At this time," observed Deputy Sheriff Spaur, "it came straight south, just one motion, buddy, just a smooth glide."

The object again turned east, flying at an estimated altitude of 300-500 feet, illuminating the ground beneath the craft as continued, seemingly pacing the squad car.

Suddenly the object turned north, passing right to left in front of the officers. Soon Spaur and Neff were doing 100 mph and above to keep up.

With dawn approaching, the men could now make out a structured, metallic form, 30-40 feet in diameter, not just a brilliant light. Spaur and Neff, in their haste, made a wrong turn — and the object turned and circled back towards the squad car, seemingly enjoying the "game" and the thrill of the chase.

By now other units on patrol became aware of the situation and came roaring from multiple directions to help pursue the aerial quarry.

Patrol Officer Wayne Huston of East Palestine, Ohio — close to Pennsylvania's western border — watched as the UFO passed overhead with the sheriff's car not far behind doing maximum speed. Huston pulled out and began following. Once in Pennsylvania, Spaur, Neff, and Huston

KECKSBURG'S SEQUEL?

pulled up to speak with an officer in a Conway, PA police cruiser. The Conway patrolman was already intently observing the UFO.

When Spaur, Neff, and Huston stopped, the UFO stopped, hovering at a safe distance. Now all four men, including the incredulous PA cop, were witnesses.

KECKSBURG'S SEQUEL?

By this time, higher authorities had been alerted, and Spaur could hear on the radio that jets were being scrambled to assess the situation. "We could see these planes coming in," said Spaur. "When they started talking about fighter planes, it was as if the thing heard every word that was said — it went PSSSCHOOO, *straight up*. And I mean when it went up, friend, it didn't play no games . . . it went **straight up!**"

If the UFO had continued on its original heading along western Pennsylvania's country highways, the cops and the saucer could've reached Westmoreland County and Kecksburg in time for breakfast.

A shaken Deputy Sheriff Dale Spaur would later make the following signed drawing of the mysterious, misshapen, inverted cone-shaped craft:

Major Hector Quintanilla, chief of Project Blue Book, is reported to have begun his "investigation" days later by asking the already on-edge Deputy Sheriff Spaur to "tell me about this mirage you saw." The interchanges between the police witnesses and Major Quintanilla were, not surprisingly, short and acrimonious.

KECKSBURG'S SEQUEL?

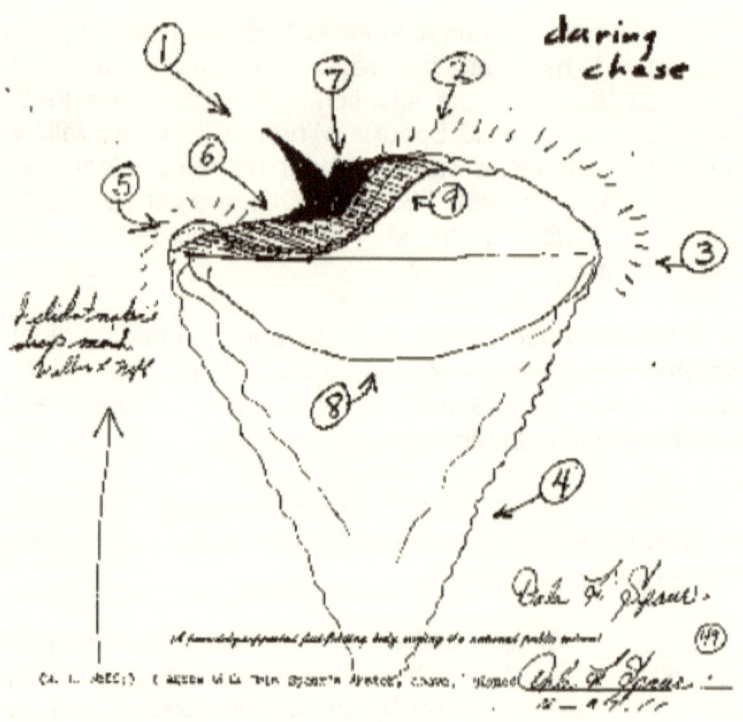

The Air Force officially ruled that what the officers had initially seen was an Echo satellite, and then later confused the planet Venus for a UFO. They also denied that any jets had been scrambled to investigate the sighting.

Deputy Sheriff Spaur would counter the official denial of what the officers had witnessed. "First of all," explained Spaur, "I don't think there is a satellite that can go this low,

KECKSBURG'S SEQUEL?

and I don't think we have one this large or that it can stop or go or maneuver any time. I'm definitely sure I wasn't chasing Venus running wildly over the countryside."

Allegedly, Spaur and Neff had the definitive proof. The officers always kept a camera in their cruiser, and took dozens of photos of the mystery UFO. But the "G-Men" from the military who investigated the incident confiscated all the negatives.

No official photos of the UFO have ever been released.

Meanwhile, Spaur and his colleagues were subsequently ridiculed as "the cops who chased Venus into Pennsylvania." They all suffered negative consequences as a result of having chased the UFO. Deputy Neff remarked to his wife that after April 17th even if the UFO were to land in the couple's back yard that he "wouldn't tell a soul."

If the Ohio-Pennsylvania UFO chase sounds vaguely familiar, it should. The April 17th, 1966 incident was the inspiration for Steven Spielberg's police-UFO chase scene in his hit movie *Close Encounters of the Third Kind*.

KECKSBURG'S SEQUEL?

Checking back a just month earlier, beginning on March 14, 1966, the State of Michigan was deluged with a spate of credible UFO sightings, from Ann Arbor to Hillsdale to Dexter. The UFO "flap" began with a couple who claimed they saw a UFO land in a marsh at the edge of their backyard. However, by the end of that day, some seven law enforcement witnesses — including police officers and sheriff's deputies from two counties — had reported seeing UFOs.

KECKSBURG'S SEQUEL?

Three days later, on March 17th, a pair of sheriff's deputies in the early morning hours saw three or four red, white, and green circular objects glowing and oscillating above the town of Milan.

Next, on March 20th, two more deputies kicked off additional sightings by following similarly described objects through Washtenaw County. Eventually six police cars would become involved in a wild UFO chase, with flashing UFOs hovering just over the road and sometimes just above a pursuing squad car. One witness who saw the chase go by his farm said the UFO looked pyramid-shaped, with "a light here and a light there and what looked like a porthole." He estimated the UFO to be the length of a car with a hazy mist that hung just beneath the craft.

On the following evening, dozens of college students witnessed, from their dorm rooms, a UFO landing next to a softball field at Hillsdale College.

The Air Force sent Project Blue Book astronomer Dr. J. Allen Hynek to investigate. His verdict? Most of the sightings were the result of "swamp gas" also known marsh gas, along with youths playing pranks with flares.

I could go on but . . . well, you get the picture. . . .

KECKSBURG'S SEQUEL?

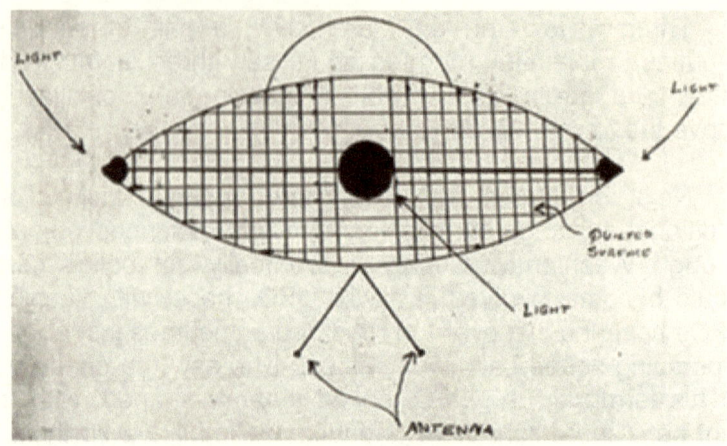

Composite drawing of the Michigan UFO

22 ♦ UFOs in the Early Days

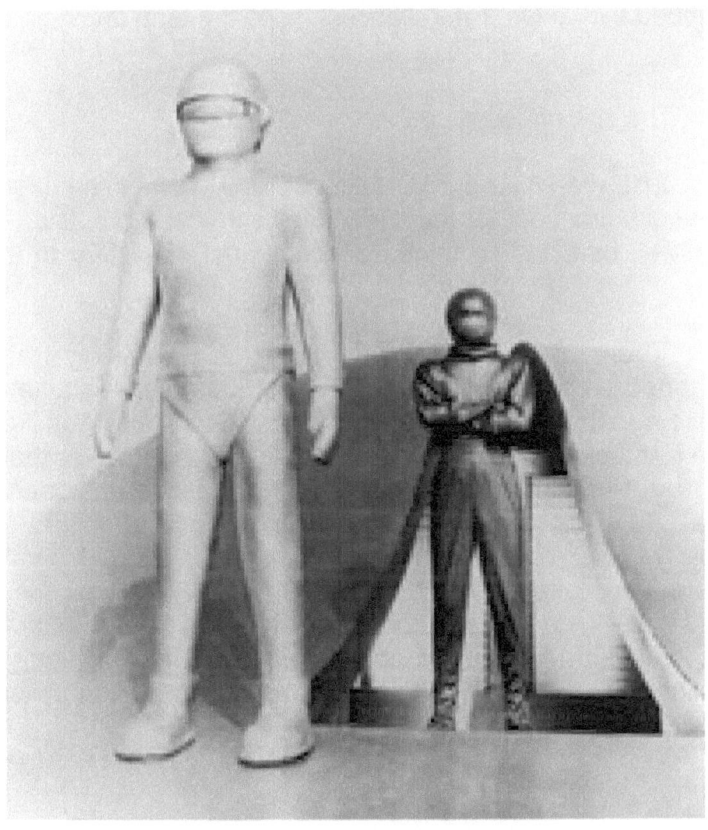

On June 24, 1947 while flying near Mount Rainier in Washington State, pilot Kenneth Arnold observed nine unusual objects, UFOs, cruising through nearby skies.

KECKSBURG'S SEQUEL?

Arnold would report his experience, describing the UFOs as "shaped like pie plates." Someone in the press quickly dubbed the UFOs "flying saucers," and the term immediately stuck.

The age of "flying saucers" had begun.

Just weeks later, in July of 1947, something highly unusual crashed in the New Mexican desert outside of Roswell that has remained a controversial mystery to this day.

Then, on January 7, 1948, Kentucky Highway Patrol reported an extremely large UFO, 250-300 feet in diameter, heading westbound. Four airborne F-51D Mustangs from the Kentucky Air National Guard were radioed to investigate. One of the pilots, Captain Thomas F. Mantell, approached the UFO, and reported back to air traffic control that the object "looks metallic and (is) of tremendous size." Mantell's plane soon after plummeted to earth near Fort Knox, killing the pilot instantly. Authorities later postulated that the experienced Mantell had actually chased either a Skyhook balloon or the planet Venus, and had blacked out at altitude from lack of oxygen.

KECKSBURG'S SEQUEL?

Two months later, in Aztec, New Mexico several witnesses claimed that a 100-foot flying saucer in distress had crash-landed on top of a mesa overlooking an oil field. U.S. military personnel arrived and quickly cordoned off the area. Allegedly the craft contained a total of 12-14 dead extraterrestrials. The *X-Files* television character Fox Mulder famously said that Aztec would prove more important than even Roswell. The fascinating, nearly forgotten Aztec case is well-covered in the book The Aztec UFO Incident by dedicated researchers Scott Ramsey, Suzanne Ramsey, and Frank Thayer PhD.

As the 1940s gave way to the 1950s, the reports of UFOs and "flying saucers" continued to escalate at an alarming rate. The popular 1951 science fiction movie

KECKSBURG'S SEQUEL?

classic, *The Day the Earth Stood Still*, about an extraterrestrial and his giant robot sidekick sent to Earth in a flying saucer to warn humanity against experimenting with nuclear weapons, most assuredly added fuel to that fire. UFO "flaps" began coming in increasingly dramatic waves.

This UFO "craze" seemed to peak in 1952. That year, according to Project Blue Book, some 1,501 sightings were logged, with more "unexplained" events than any other year in that operation's 17-year existence. Many more sightings recorded in police reports and local newspapers went officially unnoticed and unrecorded.

So, 1952 became the year of the flying saucer. Everyone was talking about them. This, of course, included Washington, D.C. and the military. The situation with the UFOs was growing critical, and quickly morphing into a national crisis.

Something had to be done. The question was . . . what to do?

KECKSBURG'S SEQUEL?

23 ♦ Earth vs. the Flying Saucers

What happened in July, 1952 finally forced the issue. It changed everything.

The July, 1952 incidents known as, alternatively, "The Invasion of Washington," "the Washington Flap," "the 1952, Washington, D.C. UFO Incident," and "the Washington National Airport Sightings," had the nation's capital on edge. Never before had Project Blue Book and the Air Force been deluged with so many sightings. UFOs were invading restricted air space with impunity. Air-traffic controllers watched in amazement when, for a period of two weeks, unauthorized objects were spotted on radar flying unconventional flight paths over and around Washington, D.C.. Not only were unidentified blips seen on radar screens, but many eye witnesses reported bright lights hovering in the sky, lights that would turn at right angles then zoom away at incredible speeds.

KECKSBURG'S SEQUEL?

Airman William Brady, from the tower at Andrews Air Force Base, witnessed "an object which appeared to be like an orange ball of fire, trailing a tail." A Capital Airlines pilot, awaiting takeoff at the National Airport, saw what he believed to be a meteor. The control tower then radioed him about six objects, fast moving lights, spotted in the vicinity. The lights eventually streaked off and subsequently disappeared from the radar tower's scope.

A master sergeant at Andrews AFB, tracking strange lights near incoming and outgoing planes, stated that "these lights did not have the characteristics of shooting stars. There was (sic) no trails . . . they travelled faster than any shooting star I have ever seen."

At one point, both radar centers at National Airport and Andrews AFB were tracking the same object hovering over a radio beacon. The object then vanished on all radar scopes at once.

Radar calculated the speeds of these UFOs at upwards of 7,000 mph.

The military and the Washington, D.C. authorities had seen enough. Air Force F-94 Starfire jet fighters were scrambled from Delaware's New Castle Air Force Base. But, their effort was to no avail, as the speedy UFOs simply toyed with the jets. Every time they angled towards the radar blips they would receive no visual confirmation. One

KECKSBURG'S SEQUEL?

wingman, Lieutenant William Patterson, reported briefly seeing four glowing balls, and attempted to give chase.

Patterson later revealed that, "I tried to make contact with the bogies below 1,000 feet. I was at my maximum speed . . . I ceased chasing them because I saw no chance of overtaking them."

At one point Patterson radioed, "I see them now and they're all around me. What should I do?"

According to ground control, nobody answered because, "we didn't know what to tell him."

Air Force brass certainly knew what to do. They called a press conference to calm an anxious public. It would be the largest Pentagon press conference held since World War II. The event was headed up by who the press would dub the Air Force's top UFO experts — Major Generals

KECKSBURG'S SEQUEL?

John Samford, USAF Director of Intelligence, and Roger Ramey, USAF Director of Operations.

The generals blamed all the sightings on misidentified aerial phenomena, namely stars and meteors. And the radar blips? They were the product of atmospheric temperature inversion, pure and simple.

Nothing to see, folks, nothing to worry about. Everything is fine, you are all safe.

Meanwhile, the Air Force was quietly issuing a directive to its fighter pilots:

Pilots Ordered to Shoot Down 'Saucers' in Range

WASHINGTON, July 28 —(INS) — The Air Force disclosed today that jet pilots are under orders to maintain a nationwide 24-hour "alert" against "flying saucers" and to shoot them down if possible.

The Air Force expressed the belief the unidentified flying objects are not a threat to the United States and stated also that they are not a secret U. S. military development.

It added, however, that jet pilots are under standing orders to pursue all unidentified flying objects, especially on the eastern seaboard, and, if necessary, force them to land. The alert is applicable to "flying saucers."

The Air Force admitted, however, that no jet pilot has yet gotten close enough to take a shot at a "flying saucer." One pilot estimated that he was within five miles of a mysterious light over Washington last weekend, but the light disappeared when he tried to draw close.

The Pentagon issued a statement

(Please Turn to Page 4 Col. 1).

KECKSBURG'S SEQUEL?

The *San Francisco Examiner* on July 29, 1952 wrote that, "The Air Force revealed today that jet pilots have been placed on 24-hour nationwide 'alert' against flying saucers with orders to '***shoot them down***' if they refuse to land.

That's right, if the UFOs couldn't be forced to land, it would be open season on the flying saucers. Shoot to kill was now the order of the day, temperature inversion or no temperature inversion.

According to UFOlogist Frank Feschino, Jr., author of Shoot Them Down! — The Flying Saucer Air Wars of 1952, this is exactly what occurred on September 12, 1952. Beginning at 3:35 a.m. and continuing for more than 18 hours, sightings of fireballs, UFOs, and flying saucers began to occur all over the eastern portion of the United States. Some of the states reporting anomalous events included Delaware, Illinois, Maryland, North Carolina, Ohio, Pennsylvania, Tennessee, Virginia, and last but not least, West Virginia. Oh, and let's not forget Washington, D.C. in this exclusive mix.

Late that afternoon of the 12th into early evening, strange things also began to happen in the skies around, yes, *Florida*. One fighter pilot out of Tyndale AFB ejected from his plane, sending his $500,000 jet (big money in 1952) crashing into the Gulf of Mexico. Next, two pilots in a Starfighter F-94 fighter jet, according to Feschino, simply disappeared off the face of the Earth. A scant paper trail was left save for the few newspaper articles that described

KECKSBURG'S SEQUEL?

the loss of the plane and pilots and subsequent futile search. The still missing and presumed dead aviators were 2nd Lt. John A Jones, Jr. of Florida and 2nd Lt. John DelCurto of Oregon. Not one piece of wreckage was ever located.

Then, as the sun started to set over the eastern seaboard on September 12th, three "meteors" began streaking in towards sensitive military locations.

One "meteor" came up from the south and, according to Frank Peschino's research of Project Blue Book files and local newspaper accounts, headed for the Oak Ridge National Laboratory in Tennessee, before turning back.

A second "fireball" came in from the Atlantic Ocean, flew over Baltimore, and proceeded towards Ohio and Wright Patterson AFB near Dayton, also before turning away.

But it was the third fireball, the one that flew west directly over Washington, D.C., that was destined to go down in the annals of the unexplained.

That third mysterious "meteor" kept on going went until it reached sleepy Braxton County, West Virginia.

Then the flaming "meteor" suddenly turned south.

Next, it *landed*.

KECKSBURG'S SEQUEL?

But meteors don't turn and "land" do they? That's because, according to Frank Feschino, the "meteor" was actually a flying saucer that had been heavily damaged under the U.S. military's "shoot to kill" orders.

And it would soon be on the ground and in trouble. . . .

24 ♦ Terror in West Virginia

The "meteor" that the *New York Times* would next day call "THE FLAME OVER WASHINGTON" passed low over D.C. at approximately 7 p.m., cruised over Virginia, and then made a beeline for central West Virginia. At approximately 7:25 the flaming object arrived over central West Virginia and the small town of Burnsville.

At that point, the "meteor" did something entirely un-meteor like. It made a 90° turn to the south.

Witnesses in Burnsville said the flaming UFO was nearly as big as a house or two-car garage, glowing or pulsating red and orange, and was definitely not a meteor. Like the Kecksburg, PA bronze-colored metallic flying acorn

KECKSBURG'S SEQUEL?

that would be seen more than 13 years later, observers would describe the Braxton County, West Virginia UFO as something that was being *flown* or *piloted* — in other words, a *craft*.

Some five minutes later, at about 7:30 p.m., and 14 miles south of Burnsville, a group of young boys playing football on a playground noticed a "fireball" zooming past a nearby mountain. Barely clearing the first mountain, the flaming UFO travelled slowly over the valley where the 300-person town of Flatwoods is located. Obviously in trouble, the glowing UFO, now decelerating, barely skimmed across the treetops of a small, flat-topped mountain on the opposing side of the valley. There it hovered for a few seconds, then dropped just out of sight behind the trees on the ridge line.

Now excited, the boys, ages 6-14, began running in the direction where they were sure the object had landed. Some were certain it was a flying saucer. Others thought it might be a meteorite, and were anxious to get their hands on some space rocks.

While sprinting towards the smallish, flat-topped mountain that sat atop the G. Bailey Fisher Farm, a few of the more timid boys began to have reservations about their safety, and so they dropped from the pack to head home.

On the way the now huffing and puffing youngsters took time to stop at the home of Mrs. Kathleen May, mother of

KECKSBURG'S SEQUEL?

two boys in the group. The May home sat close by the Bailey Fisher Farm, and the remaining six boys were anxious to tell an adult about what they had just witnessed.

Luckily for the boys, Mrs. May's cousin, 18-year old Eugene Lemon, was visiting at the May residence, and agreed to accompany the pack onto the supposed landing site. Mrs. May, still in her work uniform, grabbed her father's flashlight or electric lantern and headed out with the boys. It was now dusk and rapidly getting darker. Once on farm property, the UFO-hunters reached a grassy, uphill path that skirted some woods and the aforementioned tree-lined ridge to their left. Mrs. May instructed the boys that they would only be getting close enough to glimpse what came down before returning to the May home to notify authorities.

The uphill grassy path featured two gates, one steel and another wooden, to discourage grazing farm animals from wandering downhill and off the property. In fact, the almost perfectly flat top of the Bailey Fisher mountain was specifically used for pasture.

The UFO seekers unwired the steel gate, being careful to wire it shut behind them. Mrs. May and Gene Lemon led the way. The party proceeded, next navigating the smaller, wooden gate. While ascending towards the mountaintop, the group could see a large glowing light off to their right, now located down in a gully beside the big hill. Apparently the UFO had relocated to a less conspicuous spot. But they could still get a much better look at the situation from up top.

KECKSBURG'S SEQUEL?

Suddenly, Mrs. May smelled something strange, a curious mixture that reminded her of sulphur and burning metal. One of the boys would, years later, describe the stink as reminding him of a burnt-out radio vacuum tube. Mrs. May also noticed a warm mist, and turned to see if she could still make out the lights in town.

Just then one of the larger boys up front stopped and pointed to a large, dark figure standing to their left, partially obscured behind the trunk of a 75-foot oak tree. Two glowing eyes, about a foot apart, peered back from just under a tree limb suspended 12 feet off the ground. Mrs. May snapped on her light and pointed it at the figure, which, in her words, "lit up like a Christmas tree." Horrified, Mrs. May, Gene, and the youngsters couldn't believe their own eyes

Only 15 feet away, facing them, stood a 10-foot tall metallic green humanoid-shaped contraption that hovered about 18 inches off the path. No arms, no legs, but a pair of antennae that jutted out just below a glass-plated helmet framed within an "ace-of-spades" crown.

KECKSBURG'S SEQUEL?

In an instant the front of Mrs. May's uniform was squirted or somehow soiled with an oily black substance, and Gene Lemon caught a lungful of the noxious gas, causing him to momentarily stumble and loose his footing.

KECKSBURG'S SEQUEL?

Panic set in immediately, with everyone reversing direction and scrambling for their lives.

The terrified boys raced down the path, climbing over, under, around, and even through those two closed farm gates. Mrs. May is supposed to have leapt the lower four-foot wooden gate like an Olympic hurdler, soiled work uniform and all. Gene Lemon, retching furiously, managed somehow to stagger back to safety. Someone said they looked back and caught a peek of the green "monster" gliding left-to-right across the path towards the roiling, glowing sphere that was hunkered down in the gully.

Upon reaching the house, a frantic Mrs. May counted heads, then quickly telephoned the sheriff. Problem was, the sheriff and his deputy were already out trying to locate a reported downed aircraft and weren't immediately available.

The West Virginia State Police were notified, and the state cops authorized newspaperman A. Lee Stewart, owner of the *Braxton County Democrat*, to visit the May home and assess the situation.

Stewart found a highly agitated, red-eyed Mrs. May, the front of her work uniform heavily stained with an oily substance. Some of the boys had received bloody noses or significant knee and elbow scrapes from falling down or crashing into the gates. Several also had severely upset stomachs, especially the 18-year old Lemon, who had directly inhaled the noxious fumes.

KECKSBURG'S SEQUEL?

Stewart led a second search party up the path, although most of the boys pleaded pitifully not to have to go. Stewart and the witnesses were joined by some nearby neighbors who came heavily armed.

The posse found skid marks and areas of disturbance, but no 10-foot humanoid or huge glowing sphere. Wiser heads soon prevailed, suggesting an organized search at daybreak.

The citizens of Flatwoods never got that chance. Sometime later that night, a contingent of 30 members of the West Virginia National Guard arrived at the Bailey Fisher Farm. Led by World War II combat veteran Colonel Dale Leavitt, the guard members secured the site of the incident. In the morning they were joined by another 20-30 National Guard personnel.

Colonel Leavitt

KECKSBURG'S SEQUEL?

No downed aircraft was ever found anywhere in the county. The U.S. Air Force instructed Colonel Leavitt to obtain samples of the oily substance left on the path, as well as clippings from the oak tree and other sites around the farm. Leavitt dutifully packed these up and had the evidence sent to Washington. Leavitt never found out the results of the laboratory analysis. When asked by researcher Frank Feschino as to why the Air Force did not provide the colonel with any answers regarding the collected evidence, Colonel Leavitt, in a recorded interview, said simply, "They never do." Regarding the Flatwoods incident, Leavitt candidly admitted that "something definitely happened," and hinted at other-worldly causes.

The Air Force did, however, later publicly conclude that the Flatwoods incident was "astronomical." Basically, that the boys had seen the "Washington area meteor" that had been observed over. D.C. that same evening. As per project Blue Book, "The West Virginia Monster so-called. Actually the object was the well-known Washington area meteor of 12 Sept. landing near Flatwoods."

And the "monster" seen by Mrs. May, Eugene Lemon, and a half dozen young boys? In their excitement, they had probably mistaken a barnyard owl for an alien robot or E.T. in a spacesuit.

Next night, however, a young family driving on a lonely county road outside of nearby Frametown, WV had their automobile stall out for no apparent reason. They soon

KECKSBURG'S SEQUEL?

noticed a large glowing sphere off in the woods. When the husband went to investigate, he was stopped short by a prickly, paralyzing, electrical sensation and a horrible stench. Returning to his car with his wife and infant child, the young couple was next horrified to see a 10-foot humanoid come floating across the road to briefly examine their motionless vehicle. The green "monster" actually touched the car's fender and left burn marks.

A year later, a group of Boy Scouts camping out would be similarly terrified in — you guessed it — the State of Florida.

Newspaperman A. Lee Stewart, first on the scene at Flatwoods, had the boys each draw a picture of what they had seen. The boys' independent drawings matched almost exactly. None of the witnesses, now senior citizens, have ever waivered on their stories.

Project Blue Book cited a report by the Akron Astronomy Club which stated the "Washington area meteor" had an "actual velocity: 27 miles per second." Really? If so, then why did it take nearly 30 minutes to travel the 206 miles from Washington, D.C. to Braxton County, West Virginia? How did the Washington area meteor make a 90° turn, and fly so low and slow as to skim treetops? And most of all, why did it appear to multiple observers to be a piloted craft?

KECKSBURG'S SEQUEL?

And if it was a meteorite — a very big meteorite the size of a "two-car garage" or larger — then why was this space rock never found, despite an intensive search?

Flatwoods, Kecksburg, the Great Fireball of 1966 — we sure have a lot of "meteors" that don't act much like meteors.

With the Great Fireball of '66, the witness accounts simply do not match the official explanations for what happened. If just a meteor, why the air and helicopter search?

In Kecksburg, too many witnesses saw the Volkswagen-sized metallic bronze acorn. Unfortunately, John Murphy never lived long enough to tell the tale.

And in Flatwoods, we are arguably talking about one of the most bizarre and disturbing "close encounters of the third kind" *of all time*. The infamous incident goes by many names:

- The Flatwoods UFO
- The Flatwoods Monster
- The Braxton County Monster
- The West Virginia Monster
- The Green Monster

This amazing UFO encounter is documented in Frank Feschino's extensively researched book, <u>The Braxton County Monster</u>:

KECKSBURG'S SEQUEL?

KECKSBURG'S SEQUEL?

> Still not convinced? Here are quotes from U.S. military and intelligence sources that should give you pause:

♦♦♦

"I happen to be privileged enough to be in on the fact that we have been visited on this planet, and the UFO phenomenon is real."

— **Edgar Mitchell,** former U.S. astronaut

♦♦♦

"The Federal Bureau of Investigation has been requested to assist in the investigation of reported sightings of flying disks..."

— **J. Edgar Hoover,** former FBI Director

♦♦♦

"We have, indeed, been contacted -- perhaps even visited — by *extraterrestrial* beings. And the US government — in collusion with the other national powers of the Earth — is determined to keep this information from the general public."

— **Victor Marchetti,** former Special Assistant to the Executive Director of the CIA

KECKSBURG'S SEQUEL?

◆◆◆

"If you suppress the truth it becomes your enemy. . . . If you expose the truth, it becomes your weapon. . . . I had the evidence that a crash (in Roswell) did happen. I ask you this: were you there with me? Did you have the clearances? They can't answer these questions — they simply criticize with no evidence."

— **Col. Philip J Corso**, former Chief of the Foreign Technology Division of the United States Department of Defense

◆◆◆

"At no time, when the astronauts were in space were they alone: there was a constant surveillance by UFOs."

— **Scott Carpenter**, former U.S. astronaut

◆◆◆

"We have stacks of reports about flying saucers. We take them seriously when you consider we have lost many men and planes trying to intercept them."

— **General Benjamin Chidlaw**, formerly of U.S. Air Defense Command

KECKSBURG'S SEQUEL?

♦♦♦

"The Air Force had put out a secret order for its pilots to capture UFOs."

— **Major Donald Keyhoe**, formerly of the U.S. Air Force

♦♦♦

"These revelations underscore a long, sordid history of governmental and media secrecy and the acquisition of technologies such as microelectronics, anti-gravity propulsion and zero-point, or 'free' energy, from our visitors. This massive cover-up has been going on for almost six decades since the UFO crash near Roswell, New Mexico in July 1947, an event which was certainly not caused by balloons, as alleged by the U.S. Air force. Such myths are only accepted by the ignorant or the powerful and their subjects."

— **Dr. Brian O'Leary**, former U.S. astronaut

KECKSBURG'S SEQUEL?

◆◆◆

[Regarding Roswell, NM, July 1947] "It was not a damn weather balloon — it was what it was billed when people first reported it. It was a craft that clearly did not come from this planet, it crashed and I don't doubt for a second that the use of the word 'remains' and 'cadavers' was exactly what people were talking about."

— **Chase Brandon**, Senior CIA Office

◆◆◆

"I had a good friend at Roswell, a fellow officer. He had to be careful about what he said. But it sure wasn't a weather balloon, like the Air Force cover story. He made it clear to me what crashed was a craft of alien origin, and members of the crew were recovered."

— **Col. L. Gordon Cooper**, former Mercury and Gemini astronaut

◆◆◆

"There were actually two crashes at Roswell, which most people don't know. . . . It (the second crash) was within a few miles of where the original crash was. We think that the reason they were in there at that time was to try and recover parts and any survivors of the first crash. I'm (referring to) the people from outer space — the guys whose UFO it was."

— **Lt. Col. Richard French**, U.S. Air Force Intelligence

KECKSBURG'S SEQUEL?

♦♦♦

"An investigator for the Air Force stated that three so-called flying saucers had been recovered in New Mexico. They were described as being circular in shape with raised centers. Approximately 50 feet in diameter. Each one was occupied by three bodies of human shape but only 3 feet tall. Dressed in metallic cloth of a very fine texture. Each body was bandaged in a manner similar to the blackout suits used by speed flyers and test pilots."

— Memo from the **Washington FBI** office to **J. Edgar Hoover**, former FBI Director

♦♦♦

"You now face a new World, a world of change. We speak in strange terms, of harnessing the cosmic energy, of ultimate conflict between a united human race and the sinister forces of some other planetary galaxy. . . . The nations of the World will have to unite for the next war will be an interplanetary war. The nations of the Earth must someday make a common front against attack by people from other planets."

— **General Douglas MacArthur**, former United Nations Command and Supreme Commander, Southwest Pacific Area

KECKSBURG'S SEQUEL?

♦♦♦

Conclusion

During the 1960s as a youngster growing up in Southwest Philadelphia I was a witness to several instances of spectacular aerial phenomena:

1. As a second-grader home sick with a nasty case of mumps, while watching television on February 20, 1962, I saw legendary NASA astronaut **John Glenn** make history by becoming the first American to orbit the Earth in his Mercury capsule Friendship 7.

2. As a fifth-grader while attending a Phillies baseball game with my grandmother and sitting behind the third-base dugout in Connie Mack Stadium, I saw the Phillies' Dick Allen hit a baseball so hard it took off like a rocket, shot over the opposing shortstop's head, rose like a missile towards the left-field bleachers, and did not attain cruising altitude until it leveled off and sailed past the 15-foot Coca-Cola billboard perched high atop those same left field bleachers. The baseball left the park and presumably landed somewhere in the streets of North Philadelphia, exact location unknown. I had no idea a human being could hit a baseball so hard or so far.

3. On April 25th, 1966 I witnessed one of the largest and most amazing fireballs of the 20th century as it soared high over the City of Philadelphia. Where it came from, how high and how fast it flew, what it was, and

KECKSBURG'S SEQUEL?

where it eventually landed I cannot say. All I know for sure now is it was one of the most incredible things I've ever witnessed.

4. And on July 20th, 1969 I witnessed man's greatest technological achievement to date when astronaut **Neil Armstrong** took our species' first steps on the moon. On September 22, 1962, President John Kennedy promised the nation that, "We choose to go to the moon in this decade and do the other things, not because they are easy, but because they are hard." Less than a brief seven years later, Armstrong would report from the Moon's Sea of Tranquility that, "This is one small step for a man; one giant leap for mankind." As a newsboy for the Philadelphia

KECKSBURG'S SEQUEL?

Bulletin I still remember being the most popular person in the neighborhood on July 21st, 1969 as I was besieged for requests on the sweltering Philly streets to purchase 10¢ souvenir copies of The Evening Bulletin.

Yellow circle shows approximate spot where Allen's homerun left Connie Mack Stadium.

I believe it is safe to say that a meteorite did not land in the woods just outside of **Kecksburg**, PA on December 9th, 1965. Whatever came down to Earth was designed by intelligent beings and proved to be of immense interest to the U.S. military and NASA. Something was extricated from the woods and spirited away to a secure government facility. Possibilities include:

1. A Russian military satellite or secret weapon.

KECKSBURG'S SEQUEL?

2. A Chinese military satellite or secret weapon.
3. A U.S. military satellite or secret weapon.
4. An updated experimental version of the Nazi wunderwaffe, Die Glocke.
5. Something not of this Earth.

And the **Great Fireball of 1966**, little more than four months after Kecksburg? What was that object? Possibilities include everything that applies to the Kecksburg incident — plus also the vaguely remote choice #**6** that this was just one heck of a meteor/fireball, probably the result of a small rogue asteroid coming in contact with the Earth's atmosphere.

But why all the discrepancies as to the object's speed, height, color, direction, flight path, appearance, time of incident, whether it impacted the Earth, and, if so, where it landed?

And why did Canadian officials send in a helicopter, and also a team on the ground, if experts were so sure this was a meteor that had exploded into small fragments, pebbles, and dust nine miles above the Earth somewhere over southwestern Quebec?

Did Canadian government searchers not find anything in Quebec the way the U.S. military and NASA didn't find anything in the woods just outside Kecksburg? Or, did they find something that was kept top secret — something that in later years caused Canadian Minister of Defense Paul

KECKSBURG'S SEQUEL?

Hellyer to announce to the world that "UFO's are as real as the airplanes flying overhead."

At **Flatwoods**, something most assuredly came down on or near the Bailey Fisher farm on the night of September 12, 1952. Even the U.S. government agrees something came down, although their explanation is that it was a meteorite. But if the eyewitnesses in Burnsville and Flatwoods are correct, then the smoldering object was not a meteorite, but rather a structured craft that was piloted or remotely controlled. And if the eyewitnesses on the Bailey Fisher farm were accurate in their astounding observations, then what they saw had to be one of the following:

➢ a robot or drone acting as a sentinel for a downed alien craft

➢ a smallish alien scout craft acting as a sentinel for a downed alien craft

➢ an alien biological entity in a space suit acting as a sentinel for a downed alien craft

KECKSBURG'S SEQUEL?

Too many solid, sober, upright, expertly trained observers have seen things that they should not have seen. Many have chosen to keep silent. However, many more have come forward, even on their deathbeds, to tell what they know. Astronauts, military personnel, police, radar operators, scientists, government officials, and many others. Swamp gas, temperature inversion, flying birds, experimental aircraft, meteors, the planet Venus, and the Moon over Miami can't explain every case.

The following major incidents need to be explained. Our governments worldwide need to come clean. Something is happening in our skies — has been happening for many years — and it should be explained to the citizenry. *The time for truth is now.*

Still frame from film footage of the alleged crashed UFO near Berezovsky (Sverdlovsk region) in 1968

KECKSBURG'S SEQUEL?

The following important UFO events are just a sample of the many truly "unexplained" occurrences that cry out for investigation and explanation:

Tunguska, Russia 1908
Cape Girardeau, Missouri 1941
Roswell, New Mexico 1947
Aztec, New Mexico 1948
Washington, D.C. 1952
Flatwoods, West Virginia 1952
Kingman, Arizona 1953
Ubatuba, Brazil 1957
Antonio Villas Boas incident, Brazil 1957
Dyatlov Pass, Russia 1959
Betty/Barney Hill incident, New Hampshire 1961
Socorro, New Mexico 1964
Kecksburg, Pennsylvania 1965
Portage County, Ohio 1966
The Michigan UFO Swamp Gas Case, 1966
Shag Harbor, Canada 1967
Pascagoula, Mississippi 1973
Berwyn Mountains, Wales 1974
Coyame, Mexico 1974
Travis Walton incident, Arizona 1975
Rendlesham Forrest, England 1980
Hudson Valley, New York 1982-1986
JAL Flight 1628, Alaska 1986

KECKSBURG'S SEQUEL?

**Belgium Triangle Wave, 1989-90
Cando, Spain 1994
Phoenix Lights, Arizona 1997
Bucks County, Pennsylvania 2008-2009**

And at this time, 51 years after the fact, I would like to add the **Great Fireball of 1966** to this list.

Flaming Meteor Seen Over Phila.

About the Author

Jack Myers was born and raised in Philadelphia. He is the author of several books including *Row House Days, The Delco Files, and Treasure Kids!* Jack has worked as a teacher, newspaper reporter, newspaper editor, pizza shop proprietor, and technical writer. In his spare time Jack is very active in the pet rescue community. He currently resides in Chester County, Pennsylvania and works for a Fortune 500 financial software company.

*** If you liked this report, or have critical comments of a worthy nature, please post a review on Amazon! ***

Jack's latest book is **<u>Knights' Gold</u>**.

KECKSBURG'S SEQUEL?

Knights' Gold tells the amazing but true story of how two Baltimore boys in 1934 unearthed 5,000 gold coins hidden by a secret Confederate organization known as the Knights of the Golden Circle. These millions of dollars in gold coins represent the largest documented K.G.C. treasure find yet!

www.ingramcontent.com/pod-product-compliance
Lightning Source LLC
Chambersburg PA
CBHW021828170526
45157CB00007B/2723